Microlife

The Benefits of Bacteria

Robert Snedden

Heinemann Library
Chicago, Illinois

Produced by Paul Davies and Associates
Originated by Ambassador Litho
Printed in Hong Kong, China

04 03 02 01 00
10 9 8 7 6 5 4 3 2 1

Library of Congress Cataloging-in-Publication Data
Snedden, Robert.
    The benefits of bacteria / Robert Snedden.
      p. cm. – (Microlife)
    Includes bibliographical references and index.
    Summary: Discusses the beneficial uses of bacteria in applications ranging from brewing and baking to industry and sewage treatment.
      ISBN 1-57572-242-9 (lib. bdg.)
    1. Industrial microbiology—Juvenile literature. 2. Microbial biotechnology—Juvenile literature. [1. Microbiology. 2. Biotechnology.] I. Title.
    QR53.S572 2000
    660.6'2—dc21   99-046863

Acknowledgments
The Publishers would like to thank the following for permission to reproduce photographs:
Corbis, pp. 19, 29, 37, 41; Image Select, pp. 4, 25; Rex Features/N. Cobbing, p. 35; Science Photo Library, p. 16; M. Abbey, p. 15; M. Bond, p. 9; Dr. J. Burgess, pp. 6, 10, 13, 20; T. Buxton, p. 11; CNRI, p. 31; EM Unit/VLA, p. 7; Favre, Felix/Jerrican, p. 23; S. Fraser, p. 38; J. Heseltine, p. 24; P. & T. Leeson, p. 26; Dr. K. Lounatmaa, p. 14; Prof. P. Motta, Dept of Anatomy, University La Sapienza, Rome, p. 17; Novosti, p. 42; C. Nuridsany & M. Perennou, p. 32; A. Pasieka, pp. 5, 33; Rosenfeld Images, Ltd. p. 21; S. Stammers, p. 19; St. Mary's Hospital Medical School, p. 28; U.S. Department of Energy, p. 45; J. Walsh, p. 22; Biozentrum, University of Basel/Dr. M. Wurtz, p. 30; Still Pictures/P. Gleizes, p. 8.

Cover photograph reproduced with permission of Eye of Science, Science Photo Library.

Every effort has been made to contact copyright holders of any material reproduced in this book. Any omissions will be rectified in subsequent printings if notice is given to the publisher.

Some words are shown in bold, **like this.**
You can find out what they mean by looking in the glossary.

CONTENTS

INTRODUCTION

Think of **bacteria** and you probably think of germs, disease, and illness, and the importance of washing your hands before meals. It is true that some bacteria and other **microorganisms** cause serious illnesses, but most bacteria are not harmful. In fact, without bacteria, there would be no life on Earth as we know it.

DEATH AND DECAY

Without the benefit of the microbes in their intestines, cows would get little nourishment from the grass they eat.

Bacteria that live in soil decompose, or break down, the remains of dead plants and animals. In the process, they make available **nitrates** and other important substances used by living plants as they grow. These substances pass to the animals that eat the plants. Of course, the bacteria do not do this to be helpful. It is just the way they obtain the energy they need to survive.

WORKING ON THE INSIDE

Some bacteria colonize the digestive systems of humans and other animals. This is good from the bacteria's point of view. They get a reasonably safe home and regular deliveries of food. Coincidentally, they also perform a valuable service in exchange. Bacteria in the stomachs of grass-eating animals like cattle break down cellulose, the indigestible material that forms the stiff walls of plant **cells**. Bacteria in our intestines break down food and release **nutrients** that would otherwise pass through our bodies and be lost as waste.

DOWN IN THE MOUTH

Some kinds of bacteria even live in pits at the backs of our tongues. They get their energy by breaking down one form of nitrogen **compound** into another that can help kill the harmful bacteria that cause cavities in teeth.

MAKING USE OF MICROBES

We have found many surprising ways to turn the extraordinary properties of bacteria and other microorganisms to our advantage. We use bacteria to **ferment** milk and make cheese, butter, and yogurt. Yeasts, tiny microscopic **fungi**, are used to make bread rise and produce alcoholic drinks, such as beer and wine.

Just like larger plants and animals compete with each other for food and space so, too, do **microbes**. We can use one bacterium to combat the disease caused by another by using the weapons that bacteria have **evolved** over millions of years to gain an advantage over each other. We can even make use of the ability of bacteria to live in some of the most hostile places on the planet to run chemical processes in a more environmentally friendly way.

A *Lactococcus lactis* bacterium is one of the microorganisms that is involved in making yogurt.

No one knows how many species of bacteria there are, or what abilities they have. We may just be scratching the surface of our relationship with these marvelous microorganisms. One thing to remember is that bacteria are only doing what they have to do for their own good. If we or any other organism can benefit, it is by chance and not by design.

THE CLEAN-UP CREW

All living things produce waste in one form or another, and eventually all living things die. Imagine what the world would be like if all the waste and remains of once-living organisms were simply left to pile up. Fortunately, there is a group of organisms called decomposers. They play a vital part in the living world by feeding on this organic waste. Many of the most important decomposers are **microorganisms** such as **bacteria** and **fungi**.

DECOMPOSERS

A food chain shows the path that energy takes through an **ecosystem**. At the beginning of every food chain there are the producers, such as green plants or microscopic **algae**—which trap energy from the sun—or bacteria that obtain energy from chemical reactions. This energy can be used by other organisms, called consumers, further along the chain. They eat the producers or other consumers. The decomposers are the last link in the chain.

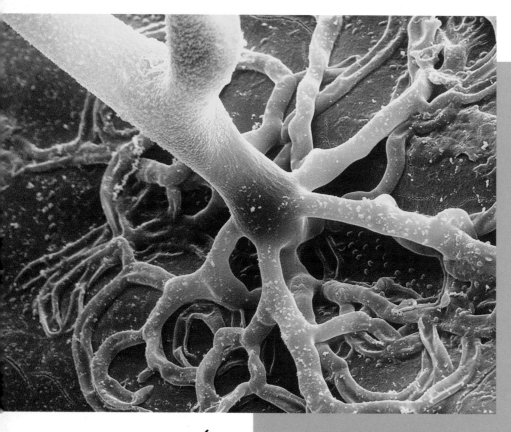

This photo shows part of a bread-mold fungus. The microscopic fungus drifts in the air until it lands on suitable food and starts to grow.

Decomposers stand apart from other consumers because they break down the food they consume into **inorganic nutrients**. This releases the nutrients back into the soil, water, and air so they can be used again. If there were no decomposers, there would eventually be no life. All of the available nutrients would become locked up in waste and dead bodies.

BREAKDOWN SQUAD

Not all decomposers are microorganisms. The earthworm is a well-known and obvious example of one of the larger decomposers. The larvae of flies and other insects also feed on dead animals. The earthworm is at the beginning of the decomposition process. It breaks down dead organic matter into smaller fragments. At this point, the microscopic decomposers—the bacteria and fungi—take over. They break the fragments down even further and release the nutrients to be used again by plants. There are billions of bacteria in every handful of fertile soil.

DECOMPOSERS IN YOU

Decomposers can be found wherever there is organic matter. There are even decomposing bacteria in our intestines that help us extract nutrients from our food. They carry on the digestive process by releasing essential nutrients from the food we eat. This is an example of a **symbiotic** relationship—one that helps both organisms. The bacteria find food and shelter in our intestines and we get vitamins and other nutrients that would not have been available to us.

Everyone has colonies of bacteria inside of them. *Escherichia coli* bacteria, shown at right, grow inside the intestines of humans.

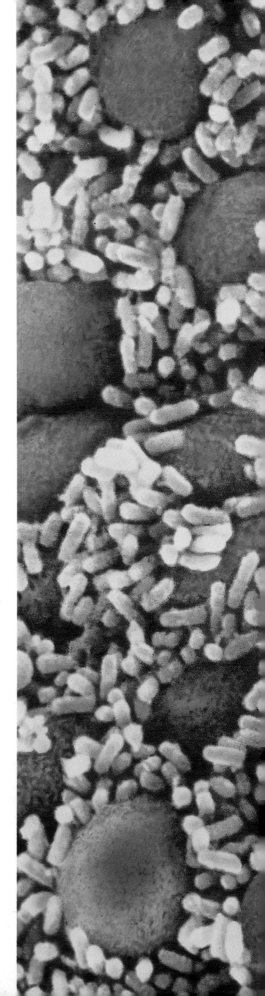

SEWAGE DISPOSAL

Every day, people on Earth produce over eleven billion pounds of human waste. Add all of the waste being produced by other animals and it amounts to a huge waste disposal problem. Luckily, we can use countless billions of **microorganisms**, including millions of different species of **protists**, **fungi**, and **bacteria**, to help face the problem.

FECES FEEDERS

Animal manure is easily broken down by microorganisms. In fact, they are at work on the manure even before it leaves the animal! Around one-third of the weight of manure is made up of bacteria such as *Escherichia coli*, a common bacterium that is found in the intestines of humans and other animals.

DIRTY WORK

As we have seen, decomposers are constantly at work recycling natural waste. However, because many of us now live in large towns and cities, we have created a waste problem that bacteria cannot solve alone. They cannot recycle the waste fast enough. We have had to find ways of working with bacteria to deal with processing our waste more efficiently.

Animal manure contains nitrogen, phosphorus, and potassium. The action of microorganisms releases these important **nutrients** into the soil.

SEPTIC TANKS

Using a septic tank is one way to treat sewage. It is usually an underground tank in which billions of **microbes** slowly break down sewage. Septic tanks contain mainly anaerobic bacteria, which live without oxygen. The process produces a great deal of gas—more than half of which may be methane. This can be collected and used as a fuel. The final solid product can be used as a fertilizer.

SPRAYING AND FILTERING

Larger quantities of sewage are treated using aerobic bacteria, which need oxygen to live. One form of treatment involves spraying the sewage onto a thick layer of crushed rock. Large numbers of microorganisms, including protists and bacteria, cling to the rock surfaces and form a **biofilm**. Some of the materials in the sewage are broken down into simpler substances by the microorganisms and large numbers of bacteria are consumed by the protists in the biofilm.

Bacteria in gravel filter beds in this sewage plant break down sewage and help to make it harmless.

A more efficient method is to use a biological **aerated** filter. This consists of a submerged bed of fine granular material, which is also coated with a biofilm. The sewage is trickled through the granular material and air is pumped in at the base of the bed. This air flow allows bacteria that break down ammonia in the wastes to grow and become established.

THE NITROGEN CYCLE

Nitrogen is essential for all forms of life. All **proteins** and the **nucleic acids**—DNA and RNA—contain nitrogen. This means that it is vital that the supply of nitrogen to the living world is maintained. The movement of nitrogen around the environment is one of the planet's great natural cycles, and **bacteria** are involved at every stage.

AN OCEAN OF NITROGEN

We are surrounded by a vast ocean of nitrogen. Seventy-eight percent of the air we breathe is nitrogen gas. However, for the majority of organisms, nitrogen is quite unusable in this form because it does not readily combine with other substances.

NITROGEN FIXATION

Some bacteria can make use of atmospheric nitrogen. Nitrogen **fixation** is the use of nitrogen gas to form nitrogen-containing **compounds**, such as ammonia and nitrate. Around 90 percent of the world's annual supply of fixed nitrogen is supplied by bacteria. A small amount of nitrogen is fixed by lightning flashes and the rest is supplied in the form of artificial fertilizers produced industrially by chemical fixation.

The presence of rhizobium bacteria, which convert atmospheric nitrogen into a form plants can use, caused these nodules to grow on the root of a pea plant.

Atmospheric nitrogen is fixed mainly by free-living soil bacteria such as **azotobacter**, by rhizobium bacteria, which associate with the legume family of plants, and by **cyanobacteria**, which are living free in soil and water and also sometimes in association with plants or **fungi**.

PLANT PARTNERSHIPS

All plants need nitrogen to ensure their growth, but none can use atmospheric nitrogen directly. Rhizobium bacteria form **symbiotic** relationships with leguminous plants, a group that includes clover, alfalfa, beans, and peas, and a few others such as alders. The bacteria live inside swellings called nodules that form on the roots of the plant in response to the presence of the bacteria. The bacteria obtain food from the plants and, in turn, supply the plants with nitrogen compounds that the plant can use to make **proteins**. Almost 200 pounds of nitrogen per acre may be fixed annually by the action of these bacteria.

ENERGY HUNGRY

Bacteria use a lot of energy to fix nitrogen. Between one-tenth and one-third of the energy a leguminous plant produces through **photosynthesis** goes to the bacteria in the root nodules.

NITROGEN RELATIONSHIPS

Many cyanobacteria can also fix atmospheric nitrogen. Some types of cyanobacteria form symbiotic relationships with fungi and plants such as ferns. For example, the cyanobacterium Anabena becomes associated with a small water fern called Azolla. Azolla is widely found, especially in the paddy fields of China and other Asian countries where rice is grown. As the rice plants grow, they shade water ferns that die and release their nitrogen for the rice plants to take up. Rice plants that grow with the water ferns give substantially better yields than those grown without—thanks to cyanobacteria.

Cyanobacteria are important in the efficient farming of rice, the staple food of almost half of the world's population.

NITROGEN CONVERSIONS

Decomposing animal remains and animal wastes contain organic nitrogen. However, this form of nitrogen is just as unusable as atmospheric nitrogen for most organisms. Many different kinds of **microorganisms** convert the nitrogen-containing waste into ammonia. Some microorganisms use the ammonia to make their own **proteins** and some of this leaks into the surrounding soil where it can be used by other **bacteria** and plants.

NITRIFICATION

Nitrosomonas and other bacteria **oxidize** the ammonia to form nitrites, and more soil bacteria such as nitrobacter oxidize the nitrites to make **nitrates**. This process is called nitrification. The nitrates can then be absorbed again by green plants. Most plants get most of their nitrogen in this form and use it to make proteins.

Nitrosomonas and nitrobacter exist in what is known as a commensal relationship. This is a relationship in which one benefits and the other neither benefits nor suffers. Nitrobacter is dependent on nitrosomonas for its supply of nitrites but nitrosomonas does not need nitrobacter.

DENITRIFYING BACTERIA

Another group of microorganisms, the denitrifying bacteria, complete the nitrogen cycle. They can convert nitrates into nitrogen gas again. This process is called denitrification and it occurs only in the absence of oxygen. It does not generally take place in well-cultivated soils. The denitrifying bacteria get their energy by breaking down both the nitrogen **compounds** excreted by living animals and the nitrogen compounds produced by decaying organic matter.

UPSETTING THE BALANCE

Human beings annually **fix** vast amounts of nitrogen for use in industry and as agricultural fertilizer. Unfortunately, such large-scale nitrogen production may be affecting the natural nitrogen cycle. There is question about whether natural denitrification by bacteria can keep pace with the amounts of nitrogen compounds being produced. If nitrates get into lakes and rivers, the result can be **eutrophication**. This is an explosive growth in the population of bacteria and other microorganisms, which thrive on nitrates. The large numbers of microorganisms use the oxygen in the water and make it uninhabitable for fish and larger organisms.

Fertilizer running into this river from surrounding fields has caused it to become choked by **algae** and bacteria.

LIVING WITH MICROBES

Billions of **bacteria** live inside human intestines. They provide an essential service by supplying essential vitamins and helping to digest food. The bacteria living in our bodies also keep harmful bacteria at bay by preventing them from gaining a foothold on "their" territory.

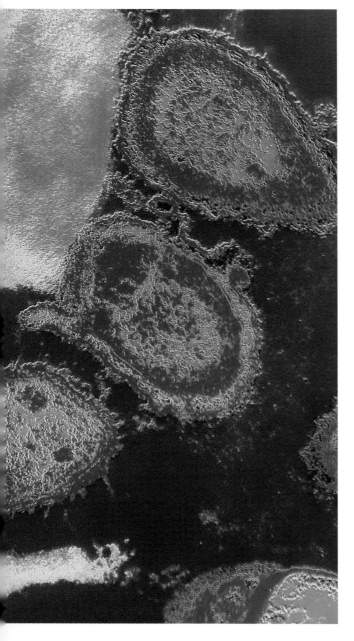

The grey material shown here is plant material in the process of being broken down by the bacteria in the rumen of a cow.

YOU ARE NOT ALONE . . .

There are more bacteria **cells** in and on your body than there are cells making up your body. In fact, your body's cells are outnumbered by one hundred to one!

MICRODIGESTERS

Bacteria and other **microorganisms** play an important part in the lives of plant-eating animals. Fifty percent of the material of a plant is made of cellulose. In fact, cellulose is the most common organic substance on the planet. However, plant-eating animals cannot digest this tough material. They need help from microorganisms.

A cow, for example, has a section of stomach called the rumen that is home to a huge range of microorganisms—hundreds of species of bacteria, forty or fifty different **protists,** and a variety of **fungi.** These microorganisms break down the plant matter that the cow consumes. The fungi break up the larger particles and the bacteria carry out the job of breaking down the cellulose into sugars. The protists digest some cellulose, but mostly they are preying on the bacteria.

TEAM TERMITE

Termites feed on wood, and some species cause serious damage in wooden structures. However, the termites cannot digest the wood. They are only able to eat wood because of the protists that live in their intestines. The protists engulf particles of wood consumed by the termite and break them down. These protists may make up as much as one-quarter of the termite's body weight. Termites deprived of their protistan partners lose the ability to survive on a diet of good wood, although they can still survive on decaying wood.

A DUAL ORGANISM

These protists are found nowhere else but in the intestines of termites. Both organisms are reliant on each other. The microorganisms can digest the wood but cannot gather it; the termites can gather the wood but cannot digest it. This is a true partnership, or **symbiosis**, where both partners gain. We could almost think of the termite and protist as a single wood-consuming organism.

Protists called *Trichonympha campanula* are organisms that inhabit the intestines of termites, where they break down the wood the termite eats.

MOUTH MICROBES

There are more than 400 different species of **microorganism**, which are mostly **bacteria,** in the mouth of every adult human. Billions and billions of them grow there on every available surface and tucked into every crack and crevice. The number of bacteria in your mouth could be greater than the number of people on Earth. Some are harmful, some are neither good nor bad, and some may actually do a lot of good.

WHAT A MOUTHFUL!

The mouth has been called the body's equivalent of a tropical rain forest because it is warm and wet and has a huge number of things living there!

RATS'S TONGUES AND ACID

Recently, scientists have discovered that bacteria living in our mouths may protect us against other potentially harmful bacteria, such as salmonella.

Experiments have shown that the tiny pits at the back of rats's tongues are home to a thriving population of *Staphylococcus sciuri* bacteria. These microorganisms take some of the oxygen from **nitrates** and use it to get energy from carbohydrates. By doing this, they convert the nitrate to nitrite. When the nitrite is swallowed, it reacts with the stomach acid to form nitric oxide, which is lethal to many bacteria and kills disease-causing bacteria in a rat's intestines.

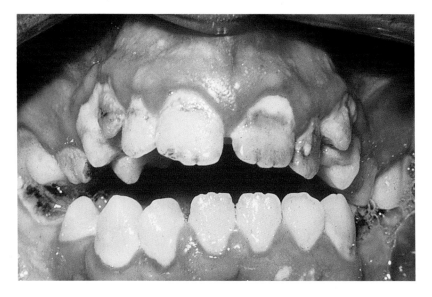

The build-up of bacteria on poorly cared-for teeth can cause tooth decay and gum disease.

The yellow objects are bacteria growing on a human tongue. Most of the bacteria in your mouth are harmless.

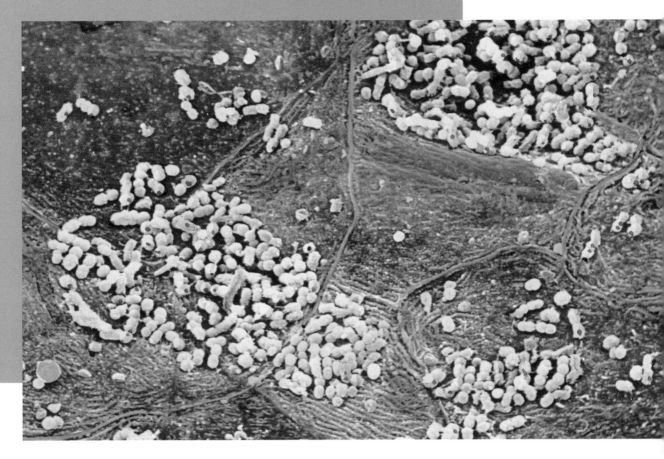

Many bacteria are not harmed by stomach acid alone. The formation of nitric oxide is a valuable additional line of defense against disease-causing **microbes**.

FIGHTING DECAY

The nitrite-forming bacteria have been found living in the mouths of humans, pigs, pigeons, rabbits, and goats, as well as rats. They probably play a part in preventing tooth decay as well. The bacteria that cause tooth decay by producing acids that dissolve tooth enamel are resistant to their own acids. However, a combination of acid and nitrite will kill them. The rural Yi, Bai, and Tai peoples of Yunnan, a province near Laos, suffer almost no tooth decay or stomach cancer. Their diet consists almost entirely of vegetables—many cured in nitrite—and small amounts of preserved meats.

FUNGAL RELATIONSHIPS

It is too easy to see **fungi** as agents of destruction. For example, dry rot is a type of fungus that can cause a lot of damage in buildings. It causes timber to become brittle and crumble into powder. Although this may be a nuisance in buildings, it is an important part of the natural recycling process. For example, when a tree is injured, it becomes infected with fungi and decays. If there were no decomposer fungi to clear the dead wood from the forests, there would eventually be no room for new trees to grow.

Fungi return organic material to the soil and keep it fertile. They also help to maintain soil structure. Some fungi produce gluelike substances that bind soil particles together. This creates **pores** in the soil and allows air and water to filter down to plant roots and soil organisms.

MYCORRHIZAL FUNGI

Of all the relationships between plants and fungi, one of the most important is the one that involves mycorrhizal fungi. Mycorrhiza means "fungus-root" and refers to the **symbiotic** relationships between certain soil fungi and the tiny feeder roots of many plants. With the exception of the growing tips, plant roots become encased in a sheath of fungal tissue. Fungal threads penetrate the plant root inside the sheath.

Fungi growing on the forest floor help to break down dead leaves and wood and recycle the nutrients they contain.

Many fungal **filaments** branch out from each mycorrhiza and absorb **nutrients** from the soil, such as phosphorus and nitrogen. These are passed on to the plant. Effectively, the fungi increase the roots's contact with the soil by 100 to 1,000 times. Mycorrhizae also give some protection from **disease agent**s and drought. In return, the fungus, which cannot produce food itself, gets carbohydrate food from the plant.

This is called the "chicken-of-the-woods" fungus. It is edible and is considered a delicacy in Germany.

FUNGI AND FORESTS

Mycorrhizal fungi are of crucial importance to the well-being of forests. The fungi absorb minerals from decaying leaves on the forest floor, which ensures that nutrients are quickly and efficiently recycled. Many of the toadstools and mushrooms you see on a walk in the woods are actually the fruiting parts of the mycorrhizal fungi associated with the trees.

ROBIN HOOD FUNGI

Scientists recently discovered that mycorrhizal fungi will actually transport nitrogen away from nitrogen "rich" plants that have nodules with Rhizobium bacteria to "poor" plants that do not. Plants that do not have bacteria to **fix** nitrogen for them can be made to grow taller and healthier by planting them next to those that do.

YEAST

Yeasts are some of the most useful **microorganisms**. These single-**celled fungi** have been used for centuries to **ferment** the sugars of rice, wheat, barley, and corn to produce alcoholic drinks.

Yeasts multiply by budding or dividing and splitting into two new cells. When placed in a sugar solution, the population of yeast will start to grow at a great rate as they consume the sugar and convert it into carbon dioxide and alcohol. Yeasts are also found in soil and salt water, where they play an important part in the decomposition of plants and **algae**.

BREAD

Yeast is used in making bread as a raising agent for dough. The most well-known and commercially important yeast is *Saccharomyces cerevisiae*. The first bread was not very much like the bread many of us eat today. It was probably flat and heavy. The bread that we eat most often is baked from fermented dough, which contains bubbles of carbon dioxide. The bubbles expand when the dough is heated. The bread rises and becomes light and airy. Left alone, dough will ferment naturally. However, this would take a long time. Therefore, a baker adds yeast to the dough to speed up the fermentation process.

Saccharomyces cerevisiae, commonly known as baker's or brewer's yeast, is used to brew beer and make bread rise.

The ancient Egyptians may have been the first to discover the use of yeast. Maybe it happened by accident when some yeast cells that were carried on the air landed on unbaked dough that had been left long enough to show an effect.

ANCIENT ALE

Microbiologist Raul Cano of the California Polytechnic State University recently succeeded in reviving a variety of microorganisms that he found in the stomach of a bee that had been fossilized in amber about 25 million years ago. Some yeast was among the microorganisms and Cano used this to brew some beer, which he said was "not too shabby." Some researchers believe it is unlikely that microorganisms could have survived so long and think that Cano's sample was contaminated with modern **microbes**.

DRUG TESTING

In early 1999, researchers at Stanford University in California said they had found a way to use yeast to screen drugs to see how they work. They hoped that this would offer a cheap and easy test for **compounds** to fight cancer, for new **antibiotics**, and other drugs.

Without yeast, bread would not rise properly, and make busy bread production lines in a commercial bakery like this one impossible. Flat breads such as tortillas are made without yeast.

DOWN AT THE DAIRY

Bacteria are used to make some of the most familiar of our foods, such as yogurt, butter, and cheese.

Yogurt is a **fermented**, slightly acidic food made from milk. Yogurt is usually made from concentrated milk that is soured by the bacterium *Lactobacillus bulgaricus*. A number of bacteria may be used in making butter and cheese.

The thin, black rods in this photograph are *Lactobacillus bulgaricus* bacteria, which are found in live yogurt.

Mass-processed yogurt is made by heating concentrated milk to about 194°F (90°C) for a few minutes, then cooling it to about 111°F (44°C) and adding a culture of *Lactobacillus bulgaricus* and *Streptococcus thermophilus*. These two **microorganisms** together produce the required acidity and flavor.

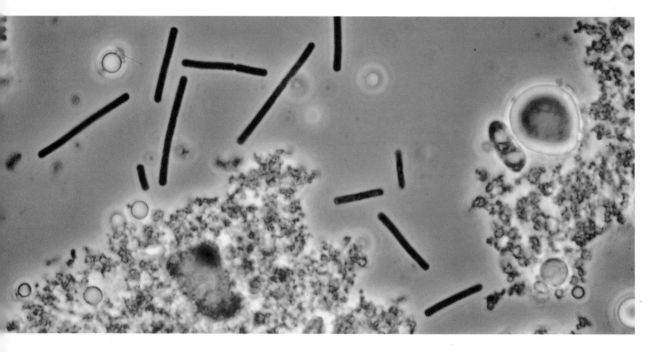

HEALTHY BACTERIA?

In the late 19th century, Elie Metchnikoff, a Russian biologist who worked at the Pasteur Institute in Paris, noticed that certain people in the Balkans, for whom yogurt was a staple part of their diet, seemed healthier and lived longer. He suggested that the numbers of harmful bacteria in the intestines were being kept down by Lactobacilli from the yogurt.

His theory was questioned by other bacteriologists who felt that conditions inside the intestine would not suit the Lactobacilli and would stop them from becoming well established.

A YOGURT CURE?

However, around one hundred years after Metchnikoff suggested his ideas, Bengt Jeppsson, Professor of Surgery at University Hospital in Malmo, Sweden, was pioneering a study of the beneficial effects of treating patients using live bacteria. The patients are given live Lactobacilli to encourage their growth in the intestines. Professor Jeppsson believes that Lactobacillus helps heal ulcers and other wounds in the bowel by keeping harmful bacteria at bay. Using bacteria like this means that the patients do not need to be given **antibiotics**, which kill off the beneficial bacteria as well as the harmful ones. Clinical trials of the new treatment were carried out in 1999.

Bacteria are used to help curdle milk in making cheese.

MAKING CHEESE

Cheese is one of the oldest and most nutritious foods we eat. Cheese-making equipment dating from 2000 B.C. has been found. It is believed that the first cheese was produced accidentally, probably through the habit of carrying milk in pouches made from animal stomachs. The bacteria in the milk and the digestive juices from the stomach worked together to first form a curd and then a crude cheese. Most cheese is formed by the **coagulation** of milk by rennet—the digestive **enzyme** in the stomach of a calf—or similar enzymes. Different types of cheese depend on the type of milk used and on the bacteria used in the ripening process— usually Lactococcus and Lactobacillus.

FERMENTATION

Certain **microorganisms**, such as yeast and some **bacteria**, can exist in the absence of oxygen. Such organisms are called anaerobes. An anaerobe is an organism that obtains all the energy it needs from food without the use of oxygen. Some bacteria not only cannot use oxygen, they are even poisoned by it. The chemical process by which living **cells** break down **glucose** in the absence of oxygen to obtain part or all of the energy they need is called **fermentation**.

FERMENTING MUSCLES

Plants and animals can also obtain energy by fermentation. When you exercise hard, your muscles cells cannot get oxygen quickly enough to obtain energy through ordinary respiration and they fall back on fermentation.

ACID OR ALCOHOL

Humans and yeast ferment glucose exactly the same way except for one small difference. When fermentation takes place in muscle cells, lactic acid is produced. The build-up of lactic acid makes your muscles feel stiff and sore after exercise. When yeast cells ferment glucose, the sugar molecules are converted to alcohol.

In a traditional process in Portugal, these people are crushing grapes to make red wine. Crushing the grape brings natural yeasts on the grape skin into contact with the grape juice.

WINE AND VINEGAR

Beer, wine, and cheese production, and several other commercial processes involve fermentation by different kinds of yeast, bacteria, and **fungus** molds. Yeasts are naturally present on grape skins. When the grapes are crushed, these come into contact with the grape sugars and fermentation begins naturally. **Cultured** yeasts are sometimes added to the mix to help fermentation.

Vinegar is produced when bacteria continue the fermentation process. They break down the alcohol produced from sugar by the fermenting yeast to give acetic acid, or vinegar. Soy sauce is fermented with the fungus *Aspergillus soyae*.

GERM THEORY

Brewer's yeast was used for brewing beer and fermenting grapes and other substances to produce wines and mashes to make distilled spirits long before the process of fermentation was scientifically understood.

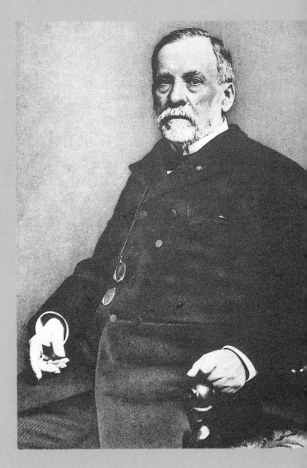

The work of the great French scientist Louis Pasteur on fermentation led to the germ theory of disease.

The French chemist Louis Pasteur (1822–1895) first demonstrated that fermentation depended on living organisms. Previously, people thought that fermentation was a purely chemical process and that microorganisms were produced by the process rather than causing it. In his *History of Lactic Acid Fermentation*, which was published in 1858, Pasteur stated that specific kinds of fermentation were caused by the activities of specific microorganisms. He also said that every disease was caused by a specific **microbe**, or germ. In this way, the study of fermentation led to the germ theory of disease.

BIOPROSPECTING

Large pharmaceutical companies screen thousands of **microbes** a week to search for new chemicals to use in the fight against disease. It can be a long search. On average, one useful new drug is found for every 20,000 natural **compounds** that are tested. However, the bioprospector—as they are sometimes called—come up with something often enough to make the search worthwhile.

TREE TREATMENT

The Pacific yew tree is an important natural source of a potent drug that is used to treat certain cancers, such as breast cancer. The tree actually produces the substance to combat cancer cells. The amount of the drug that can be obtained from each tree is too small to meet demand and obtaining it kills the tree. It is possible to make the drug chemically, but this is expensive. However, a **fungus** has been found that grows on the Pacific yew and produces the drug itself.

Maybe thirty to fifty different fungi grow on a single limb of the yew tree. Many of these microbes, called endophytes, receive **nutrients** from the tree and, in turn, give it protection from disease-carrying **bacteria** or fungi. One of these fungi appears to make the anti-cancer chemical on its own, even when grown apart from the tree. This ability probably gives it an advantage in competition with other microbes. Growing the fungus is an environmentally sound and relatively simple and inexpensive way to produce the anti-cancer drug.

FUNGUS FLAVORING

One of the first chemicals to be mass produced from fungi was citric acid, which is now used to flavor soft drinks and foods. The acid was originally extracted from citrus fruit, such as lemons and limes. In 1917, however, a chemist named James Currie realized that it could be made more cheaply using fungi.

SEARCHING FOR NATURAL CURES

Researchers are interested in discovering other plant-associated microbes that make compounds with medicinal value. For example, the National Cancer Institute collects 1,000 samples of fungi and **algae** each year, as well as 1,000 marine invertebrates and 3,000 plant samples from tropical rain forests throughout the world.

BURIED TREASURE

Ivermectin is a product of a soil bacterium. It can combat **parasitic** worm infections in humans and animals better than any other known substance. It paralyzes the worms by blocking the transmission of nerve signals. A drug used to lower levels of **cholesterol** by blocking the body's production of this substance was also found in a soil sample. There might literally be wonders in the dirt beneath our feet.

The trunk of a Pacific yew tree. The anti-cancer drug Taxol can be extracted from the bark of the tree. Because the tree is scarce, scientists are looking for other ways to obtain Taxol, perhaps from fungi that grow on the tree.

27

ANTIBIOTICS

The first clue that drugs might combat illness using microscopic organisms emerged in 1928. Alexander Fleming noticed that a mold had invaded a petri dish upon which he was growing a **culture** of **bacteria**. The bacteria had disappeared where the specks of green mold had appeared on the dish, as shown above.

AGAINST LIFE

The effect the mold had on the bacteria was *antibiosis*, a word coined in the 19th century meaning "against life," and used to describe a type of natural competition among species. Fleming tested the mold on a range of bacteria and found that it killed some disease bacteria, but not all of them. He identified the mold as *Penicillium notatum*, a similar mold to that which grows on stale bread, and named the substance it produced "penicillin."

FOLK WISDOM?

A traditional folk remedy for dealing with infections was to place a piece of moldy bread on the affected body part.

THE ANTIBIOTIC ERA

Dr. Selman Waksman, a Russian-born American microbiologist, first used the term "**antibiotic**" in 1945. The discovery of penicillin was the beginning of the antibiotic era in which pharmacists around the world searched for newer and more effective antibiotics. Today doctors can choose from almost one hundred antibiotics. Some are known as broad-spectrum antibiotics, which means that they are effective against a wide range of bacteria. Others are specific to a single strain of bacteria.

FIGHTING RESISTANCE

Increasingly, bacteria have become resistant to the antibiotics we use. Drugs that easily destroyed bacteria in the past can no longer do the job. Common bacteria such as Staphylococcus are resistant to the most powerful antibiotics we have.

Different strains of Staphylococcus bacteria are being grown in the laboratory to test their resistance to various antibiotics.

STREPTOGRAMINS

Synercid is the first of a new type of antibiotics called streptogramins. The streptogramins are the first new class of antibiotics to be introduced in twenty years. The first one was discovered by researchers searching for new medicinal **compounds** in a soil sample in Argentina. Because earlier antibiotics had been so effective and because the new drug was difficult to make in any quantity, it was ignored for years. However, now methods have been developed to produce streptogramins on a commercial scale.

29

ONE AGAINST THE OTHER

The increasing resistance of disease-causing **bacteria** to **antibiotics** is a big problem, and medical science must find an answer. One possible solution might lie in using **bacteriophages,** which are **viruses** that attack bacteria.

DISCOVERING BACTERIOPHAGES

In 1917, Felix d'Herelle, a bacteriologist at the Pasteur Institute in Paris, was investigating an outbreak of the disease dysentery when he found something that attacked the dysentery bacteria. He believed it was a virus that was **parasitic** on bacteria. He called the virus a bacteriophage.

He suspected he might have discovered something that could be used to fight infection. At first, he was scarcely believed because it seemed extraordinary that microscopic bacteria should themselves be infected by something even tinier. In the 1930s, bacteriophages were actually sold for treatments of dysentery, typhoid, and other illnesses. However, the results were unconvincing and the rise of **antibiotics** soon pushed bacteriophage treatment to the background.

T4 viruses, like this one shown here, specifically attack *E.coli* bacteria and are found in sewage and polluted water.

MINIATURE ATTACKERS

A typical bacterium is about a thousandth of a millimeter across, and a bacteriophage is about forty times smaller than that. Today, we have electron microscopes, a tool not available to d'Herelle, that can reveal them to us. They look like microscopic lunar landers with angular heads, long tails, and spindly legs that they use to cling to the surface of a bacterium. When a bacteriophage makes contact with a suitable bacterium, it uses its tail to make a channel through which it shoots its **genes** from its head into the inside of the bacterium. There, the viral genes take control of the bacterium by turning it over to the production of more viruses. Within a short time, a fresh fleet of viruses burst from the bacterium. The bacterium is left damaged beyond repair and soon dies.

FUTURE TREATMENTS

Today, antibiotic-resistant bacteria such as Staphylococcus, a common cause of infections in hospitals; Streptococcus, the cause of scarlet fever; pneumonia and the so-called "flesh eating" infections; and the tuberculosis bacterium, have turned scientists' attention again to the use of bacteriophages. There could be real advantages if they were made to work. Unlike antibiotics, bacteriophages only target specific bacteria. Antibiotics can often wipe out good bacteria as well as bad. However, the bacteriophage is so specific that it is essential to match virus and bacterium exactly when carrying out treatment. Presently, bacteriophages have not been found for all bacterial infections.

This photo shows *Mycobacterium tuberculosis*, the bacteria responsible for tuberculosis. Finding a bacteriophage that can attack this bacterium would be a great benefit.

31

BIOLOGICAL CONTROL

Biological control means using one type of organism to control the numbers or activities of another. It takes advantage of the natural competition between organisms, even at a microscopic scale, to combat pests and diseases that affect domestic animals and crops.

THE SAFER SOLUTION

The advantage of biological control is that it tends to be safer than the alternative—the use of chemical pesticides. One way to use **microorganisms**, or the **toxins** they produce, is to kill or discourage insect pests of crop plants. The **bacterium** *Bacillus thuringiensis* is an important bioinsecticide that is employed all over the world against pests that affect a variety of crops. Some strains of this bacterium produce toxins that will kill mosquitoes, and efforts are being made to develop an insecticide that can be used to combat diseases spread by mosquitoes, such as malaria and yellow fever.

One problem with biocontrols is in finding organisms that are robust enough to be taken to where they are needed and then multiply enough to do the job efficiently. If possible, it is best to recruit an organism that is native to the area being treated.

The bioinsecticides include baculoviruses— a group of viruses that cause disease in caterpillars, a major agricultural pest.

Taking the Toxins

Genetic engineers have created new strains of crop plants with built-in insecticides by inserting bacterial toxin **genes** into the plant **chromosomes**. This has the advantage of keeping insect pests under control without exposing other plants and wildlife to insecticide sprays. Once again, the plant genetic engineers have used toxins from *Bacillus thuringiensis*. There are concerns that the insects will eventually become resistant to the toxins, so scientists continue to look for other possibilities. One group of toxins that might eventually be used comes from the bacterium *Photorhabdus luminescens*. This bacterium produces **proteins** that are effective against a wide range of common insect pests.

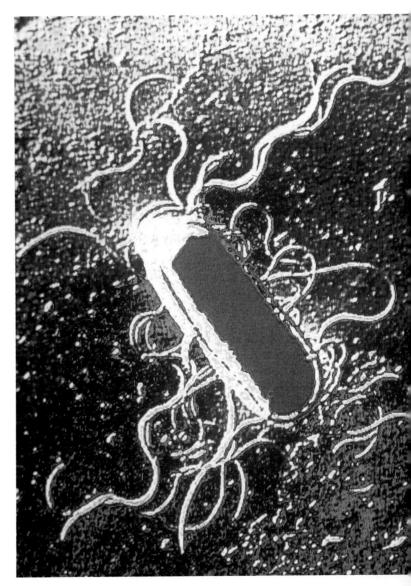

Microencapsulation

A new technology called microencapsulation could help useful **viruses**, bacteria, and other environmentally friendly biopesticides compete with traditional chemical pesticides. Encapsulation involves mixing the **microbes** with a material such as cornstarch that has been treated to enable it to absorb water. The cornstarch-microbe mixture is added to water and then dried. The microbes become entrapped in protective particles so small they can barely be seen. The biopesticides can be dusted onto crops.

Because *Bacillus thuringiensis* is harmless to man, it can be used on crops close to harvest time when it would be too dangerous to use chemical pesticides.

33

GENETIC ENGINEERING

Genetic engineering uses biochemical techniques to change the hereditary material, or genes of cells. Genes are made of deoxyribonucleic acid, or DNA, and genetic engineering includes snipping out bits of DNA and transferring them from one cell to another. This allows changes to be made to the characteristics of an organism that could not be achieved by simple cross-breeding.

PLASMID TRANSFERS

Most of the work in transferring genes has been done with bacteria. The genetic information of a bacterium is contained in a single, large DNA molecule, its chromosome. Small circular segments of DNA, separate from the main chromosome, are also found in bacteria. These are called plasmids. It is possible to take plasmids from one cell and introduce them into another. For example, *Escherichia coli*, a common bacterium found in the digestive systems of many animals, contains a plasmid that gives it resistance to the antibiotic tetracycline. *E. coli* cells that lack this resistance can be treated chemically to make them take up plasmids from other cells. In this way, they also become resistant.

ENZYME SCISSORS

It is not possible to slice off pieces of DNA using a conventional blade. Instead, it is done by using various enzymes known as nucleases or restriction enzymes. These are produced by bacteria as a defense against bacteriophages, the viruses that attack them. The bacteria fight back by using nucleases to cut up the viral DNA. Scientists can use these enzymes to break up DNA molecules.

The pieces of DNA are then joined using a type of enzyme called a ligase that repairs breaks in DNA. Scientists used nucleases and ligases to make hybrid plasmids by joining pieces of plasmid from two different bacteria species.

34

Not everyone agrees that genetic engineering is good. Greenpeace has argued against genetically modified, or GM, food with their "True Food" campaign. Some people have resorted to direct action by destroying GM crops.

BACTOADIUM?

Genes from a toad were the first animal genes to be introduced into bacteria. They were joined to plasmids from *E. coli*. The clones that resulted had toad DNA as well as that of the bacterium.

TRANSGENIC ORGANISMS

It is now possible to transfer genes from one animal to another, or from an animal to a plant. The DNA may be injected directly into a fertilized egg, or a virus may be used to carry the gene with it as it invades a cell. Human genes have been transferred to plants such as tobacco in experimental efforts aimed at producing large amounts of antibodies and enzymes to be used medically. A transgenic organism is one that contains genes from two different species.

Interferon is a naturally produced **protein** that fights viral infections. A single quart (about one liter) of bacterial culture that has had the gene for interferon inserted can produce as much of the protein as thousands of quarts of blood in the human body. Bacteria-produced insulin has also become important in the treatment of diabetes. In agriculture, applications of genetic engineering include the development of plants with resistance to insect pests and disease.

35

GENE THERAPY

Human beings have around 100,000 different **genes**. Each one of them is an instruction coded in **DNA** to produce **proteins**, the molecules that help organize and direct every aspect of the way the body works. If a gene is defective, the consequences can be disastrous. About 4,000 genetic diseases have been identified. They are passed down from parent to child, generation after generation. Until recently, they have never been considered curable, but this may soon change.

VIRAL MESSENGERS

Gene therapy is a new branch of medicine that has been made possible by genetic engineering. Methods have been developed to take normal copies of human genes and give them to people who have inherited damaged genes. The "messengers" used to carry the healthy genes into the patient's **cells** are **viruses**.

Viruses are a very good way of delivering genes to cells. It is easy to insert an additional gene into the viral DNA, which produces a **hybrid** virus called a recombinant delivery vector. The virus itself is made harmless so that it cannot cause disease, but the genes it carries will be read and used as if they were part of the host cell's own DNA. This technique involves taking white blood cells from a person with a genetic disease and introducing a normal gene into the defective cell. The normal gene is delivered using a retrovirus—a type of virus that inserts its genes into the DNA of the target cell—and introduces the properly functioning gene as it does so.

DEVELOPING THERAPIES

Gene therapy is now being developed for the treatment of nearly twenty **hereditary** diseases, including cystic fibrosis. In this illness, a defective gene causes the production of mucus that is too thick to carry out its normal function of removing particles and **bacteria** from the airways. The thickened mucus clogs the airways and makes breathing difficult.

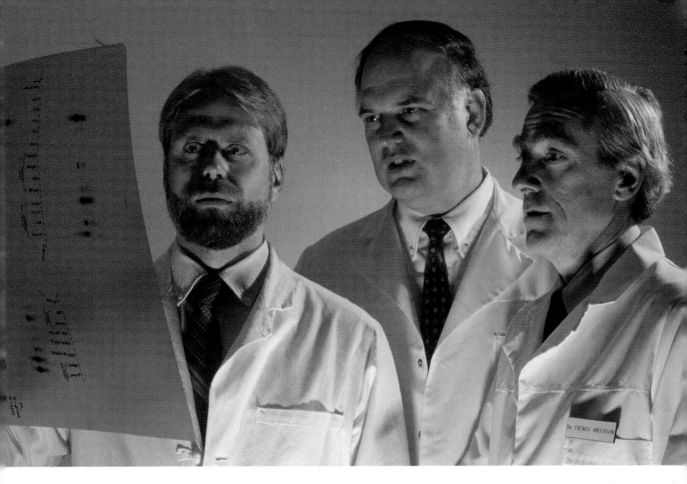

For cystic fibrosis, another kind of virus called an adenovirus has been used as the carrier for the healthy gene.

ASHANTHI DESILVA

The first human genetic disease to be treated by gene therapy was a form of severe combined immunodeficiency disease, or SCID. This illness affects children who are born without a type of white blood cell called a T cell. Without T cells, the immune system, our protection against disease, cannot work properly. One form of SCID is caused by a defective gene. In 1990, a team of researchers at the National Institutes of Health used gene therapy to treat four-year-old Ashanthi DeSilva. Some of Ashanthi's cells were removed from her blood and exposed to the virus containing the healthy gene. The viruses, which had been made harmless, invaded the cells and carried the gene with them into the blood cell DNA. They were then returned to her body. After a few treatments, her immune system was almost completely normal.

Doctors Culver, Blaese, and French Anderson (from left to right), carried out the first successful treatment of a genetic disease using gene therapy.

Extreme Enzymes

Many industrial processes are carried out at high temperatures and pressures and involve the use of substances that are harmful to the environment. Biological substances, because they can be broken down naturally, are considered to be more environmentally "friendly," but their use in industrial processes has been limited by their inability to withstand the harsh conditions involved.

Extremophiles

Some **microorganisms** can flourish in surprisingly hostile environments. For example, some are found around undersea volcanic vents where the water temperature approaches the boiling point. These are the extremophiles, members of a unique kingdom of **bacteria** called the Archaebacteria. The extremophiles represent the most ancient forms of life on Earth. They appear to have **evolved** at a time when conditions on our planet were much harsher than they are now.

Dissolved minerals and the bacteria that live in the Morning Glory hot spring in Yellowstone National Park produce colors. Some of the **cyanobacteria** can tolerate temperatures of 330° F (166° C).

Extremozymes

The robust **enzymes**, or biological **catalysts**, that these bacteria employ in their essential life processes are sometimes called extreme enzymes or extremozymes. The unique properties of the extremozymes make them an attractive proposition for use in industry. For example, a bleaching enzyme produced by an extremophile found in the scalding springs of Yellowstone National Park could provide an alternative to chlorine for whitening paper.

A heat-tolerant extremozyme in commercial use has helped to increase the production of **compounds** called cyclodextrins from cornstarch. These have many uses including improving the body's uptake of medicines and reducing bitterness and concealing unpleasant odors in food and medicines.

COLD LOVERS

Extremophiles are found in very cold places, too. These are called psychrophiles, or cold lovers. Manufacturers who need enzymes that work at low temperatures have become interested in psychrophile enzymes. They would be useful in industries such as food processing, where temperatures are kept low to avoid spoilage and in the manufacture of perfumes, which evaporate at high temperature.

BIOLOGICAL WASHING

Alkali-tolerant extremophiles are found in soda lakes in such places as Egypt and the western United States. Detergent makers are particularly interested in their enzymes. In detergents, proteases—which break down **proteins**—and lipases —which break down grease—must effectively cope with food stains. However, detergents tend to be highly alkaline, and this destroys ordinary proteases and lipases. Extremophile versions of these enzymes that can operate efficiently in heat or cold are now in use or being developed.

HOME-GROWN EXTREMOZYMES

It is difficult to obtain extremozymes in sufficient quantities. One approach is to use **genetic** engineering to insert the gene for an extremozyme into a more easily grown bacterium, such as *Escherichia coli*. This eliminates the need to obtain the bacteria from their extreme locations.

BIOMINING

Biomining uses **microorganisms** that dissolve minerals out of rocks. Through this method, minerals can be obtained without the need for heavy machinery used in other mining methods. Recovery rates have increased and operating costs decreased where biomining has been used.

MICROBE MINERS

Biomining is not a new discovery. Two thousand years ago, the Romans noticed that water running off the waste pile of one of their copper mines in Spain was blue with copper salts. They found a way to recover the copper but had no idea as to how the metal had been dissolved.

In the mid-20th century, the **bacterium** *Thiobacillus ferrooxidans* was found to be responsible. This bacterium gets energy by combining **inorganic** materials—such as sulphide-containing minerals—with oxygen. As part of their natural chemical processes, the bacteria release acids and other chemicals, which can wash metals out of ore.

ACID STARTER

Biomining is an inexpensive way to extract copper from ores where the metal is found in rocks that contain sulphur. The bacteria attack the ore, which is treated with sulphuric acid to get the process off to a good start—like adding a starter to a compost pile. As the bacteria take the **nutrients** they need from the sulphur **compounds**, the copper is released and concentrated in a solution that is collected. The metal is extracted from the solution, which is then recycled and used to prime the next batch of biominers.

BOOMING BIOMINING

The copper industry was quick to put biomining to work. About 25 percent of world copper production is based on biomining.

GOLD DIGGERS

Thiobacillus ferrooxidans is also being used successfully on gold-bearing ores. Low-grade gold ore often contains the metal bound up with sulphides such as copper ore. Conventional extraction methods require roasting or high-pressure **oxidation** to burn off the sulphides before the gold is then extracted using cyanide. Using bacteria not only removes the need for these costly procedures, it can sometimes nearly double the rate of gold extraction.

ENGINEERING BIOMINERS

In the future, scientists may look at ways to genetically alter the bacteria to make them even more efficient. Certainly, there will be objections to releasing genetically modified organisms into the environment in this way.

Raw copper ore is ground into slurry here before bacteria are used to separate the copper from the rock.

OIL WELLS, BIOFILMS, AND UMBS

Bacteria often form colonies called **biofilms**. A biofilm may contain billions upon billions of **microorganisms** attached to a surface, which could be a rock, a piece of machinery, or even a tooth, as long as there is a plentiful supply of food. The bacteria produce a slimy coat that not only gives protection but also acts as a trap for food.

ULTRA-MICRO BACTERIUM

The other side of the bacterium's lifestyle is the ultra-micro bacterium, or UMB. When food is scarce, a bacterium can shut down most of its activities and shrink to about one-third of its normal size. It can remain in this **dormant** state, maybe for centuries, until conditions improve. If conditions sufficiently improve, the dormant bacteria can form biofilms again.

OIL EXTRACTION

The oil industry is always looking to extract more oil from its wells. When an oil well is first drilled, about one-third of the oil deposit will gush out simply because it is under pressure. Perhaps one-third more can be recovered by pumping water into the well to increase the pressure. Obtaining more oil than that is difficult because the injected water leaks into rocks around the oil deposit, and oil gets stuck in the fine tubes and holes of the **porous** rocks.

The Earth's deposits of oil and other minerals will not last forever. Biofilms might enable us to make the most efficient use of these precious resources.

LIFE IN THE DEPTHS

Scientists believe that the ability of bacteria to change between biofilms and UMBs may be put to good use. One way to plug the holes in the rocks around oil deposits is to use a biofilm. In the early 1990s, the U.S. Department of Energy sampled for **microbes** while drilling for oil 8,005 feet (2,440 meters) below the surface. The researchers found that there was a surprisingly rich variety of bacteria in the oil wells.

TINY SPHERES

All bacteria will go into the ultra-micro form when they are deprived of food. About 60 percent become tiny smooth spheres. When faced with a leaky oil well, scientists can place samples of the bacteria in the well and look for those that form smooth shapes. These bacteria are then grown in the laboratory until they reach a very high concentration, about 500 billion bacteria per milliliter. Then they are deprived of food, which causes them to enter the ultra-micro state. In this form, they are shipped back to the oil well as a paste.

BACTERIA IN THE BOREHOLE

At the well, the bacteria paste is diluted and injected into the borehole. A rich food supply is then injected after them. The bacteria emerge rapidly from their dormant state and form a biofilm on the rocks that will block all the holes and prevent leakage.

Dr. Bill Costerton developed this procedure at Montana State University. He thinks that his approach can be applied more generally, perhaps to contain polluted sites or to make harmful substances safe.

ANTI-POLLUTION MICROBES

Some types of groundwater-living **microbes** have the ability to consume and digest polluting chemicals. The **microorganisms** use the pollutants as a source of food and energy by breaking them down in the process into byproducts such as water, carbon dioxide, methane, and hydrogen. This can bring the level of contamination back down to safe levels again.

MAKING A LIVING

Microorganisms such as **bacteria** have had millions of years and countless millions of generations in which to **evolve** a host of complex biochemical ways to live from seemingly unlikely materials. Huge, diverse populations of bacteria exist below the earth's surface that have the ability to break down many materials which are both natural and artificial.

BIOREMEDIATION

Bioremediation is the use of biotechnology to clean up pollutants from the environment. Bioremediation technology can be applied to polluted ground without any soil being excavated. First, appropriate strains of bacteria that can be found in the contaminated area need to be identified. The bacteria are then **cultured** in a laboratory to bring their numbers up to the required level. The microorganisms are then re-injected into the soil in the polluted area where they can get to work breaking down the pollutants.

THE RIGHT BACTERIA FOR THE JOB

One problem with bioremediation is that finding and successfully culturing suitable bacteria can be costly and time consuming. It also may take years for the bacteria to clean up the polluted area. Scientists are looking at ways to extract the **enzymes** used by the bacteria to break down the pollutants. If these could be successfully produced in the laboratory, it might be possible to treat the pollution problem directly without the expense of maintaining a large microbe population.

A laboratory technician adds microbes to a solution of toxic chemicals to observe how well the microbes break down the chemicals.

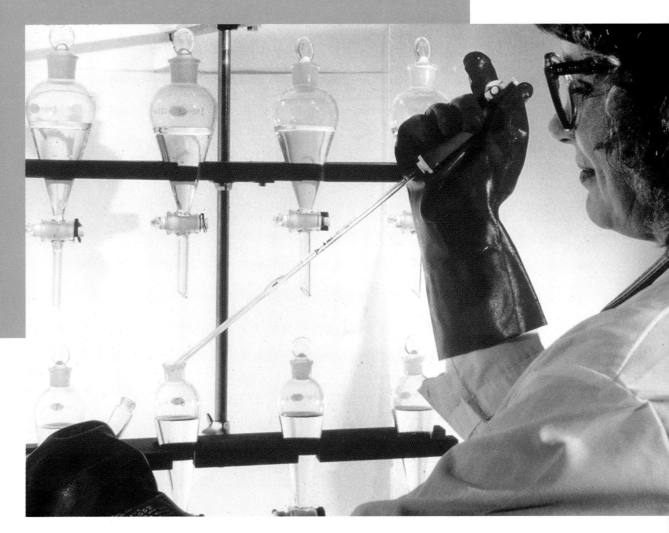

BETTER BIODIGESTERS?

Bacteria can even take on chemical pollutants, such as polychlorinated biphenyls, that resist other treatment. Scientists hope to develop strains of bacteria that will biodigest multiple chemicals because most polluted areas are contaminated with more than one chemical. However, there would inevitably be resistance to the idea of introducing genetically engineered bacteria into the environment.

GLOSSARY

aerated supplied with a gas

alga(e) simple form of plant life, usually living in water, ranging from a single cell to a huge seaweed

antibiotic substance produced by or obtained from certain bacteria or fungi that can be used to kill or inhibit the growth of disease-causing microorganisms

azotobacter large, rod-shaped bacteria that occur in soil or sewage

bacteriophage virus that attacks a bacterium

bacterium (plural bacteria) any of a large group of single-celled organisms that have no organized nucleus

biofilm colony of billions of bacteria living on the surface of something that can provide them with water and nutrients, producing a protective, slimy coat

catalyst something that can change the rate of a chemical reaction without itself being altered in the reaction

cell basic unit of life, existing as independent life forms, such as bacteria and protists, or forming tissues in more complicated life forms, such as muscle cells and nerve cells in animals

cholesterol substance found in the body that is an essential part of cell membranes

chromosome threadlike structure that becomes visible in the nucleus of a cell just before it divides, and that carries the genes that determine the characteristics of an organism

coagulation transformation of a liquid into a soft, semi-solid mass

compound substance formed from two or more chemical elements

culture microorganisms grown in the laboratory in a nutrient substance

cyanobacterium type of bacterium that is capable of photosynthesizing

disease agent organism that can cause disease in another organism

DNA (deoxyribonucleic acid) genetic material of almost all living things with the exception of some viruses, consisting of two long chains of nucleotides joined together in a double helix

dormant temporarily inactive but able to be activated again at some point

ecosystem plants and animals in an area and the environment that affects them

enzyme type of protein that acts as a catalyst, altering the rate of a biochemical reaction

eutrophication process by which microorganisms, such as bacteria and algae, grow in large numbers in water that is rich in nutrients, resulting in the death of other organisms as the microorganisms use up the oxygen in the water

evolve in biology, to develop a characteristic over a period of time as a result of mutation and natural selection

ferment to cause the chemical breakdown of sugars using bacteria in the absence of oxygen

filament fine, thread-like structure

fixation the act of fixing. In biology, nitrogen fixing is the process by which nitrogen is converted into compounds that can be used by living things.

fungi group of spore-producing organisms that includes mushrooms and molds

gene   length of DNA carried on a chromosome that acts as the unit of heredity, containing the set of instructions for assembling a protein from amino acids

glucose   simplest form of sugar, produced by green plants and some other organisms by combining carbon dioxide and water in photosynthesis

heredity   passing of characteristics from parent to offspring by means of the genes that the offspring inherits

hybrid   something that results from the combination of genetic material from two different species

inorganic   describes something that is not produced by living things

microbe   another name for a microorganism

microorganism   any microscopic living thing, such as bacteria and protists

nitrates   compounds of nitrogen in which each nitrogen atom is combined with three atoms of oxygen

nucleic acids   (DNA and RNA) DNA encodes genetic information and RNA "reads" this information and translates it into protein production.

nutrient   nutritious substance found in food

oxidize   to combine a substance with oxygen

parasite   organism living on another and benefiting without giving anything in return

photosynthesis   process by which green plants and some microorganisms make carbohydrates from carbon dioxide and water using the energy of sunlight

plasmid   circular strand of DNA found in bacteria that is separate from the main chromosomal DNA

pores   minute openings, or tiny spaces in rock or soil

protein   one of a group of complex organic molecules that perform a variety of essential tasks in living things, including providing structure and controlling the rates of chemical reactions

protist   single-celled organism that is a member of the kingdom Protista

RNA   (ribonucleic acid) found in different forms within cells, involved in the process by which the genetic code of DNA is translated into the production of proteins in the cell

symbiotic   describes a close association between organisms of two different species that is of benefit to both

toxin   poisonous substance produced by an organism such as a bacterium

virus   infective particle, usually consisting of a molecule of nucleic acid in a protein coat

MORE BOOKS TO READ

Facklam, Howard, and Margery Facklam. *Bacteria*. Brookfield, Conn.: Twenty-First Century Books, 1995.

Facklam, Howard, and Margery Facklam. *Viruses*. Brookfield, Conn.: Twenty-First Century Books, 1999.

Pascoe, Elaine. *Slime, Molds & Fungi*. Woodbridge, Conn.: Blackbirch Press, 1999.

INDEX